SEA STARS

Rebecca Woodbury, Ph.D., M.Ed.

Gravitas Publications Inc.

SEA STARS

Illustrations: Janet Moneymaker

Sea Stars
ISBN 978-1-950415-62-5

Published by Gravitas Publications Inc.
Imprint: Real Science-4-Kids
www.gravitaspublications.com
www.realscience4kids.com

RS4K

Photo credits: Cover & Title Pg: By dam, AdobeStock; Above: Image by simone saponetto from Pixabay; P.7: Photographer-Claire Fackler, CINMS, NOAA; P.9: Milada Vigerova from Pixabay; P.10: 1. Sophia Hilmar from Pixabay; 2. Simone Saponetto from Pixabay; 3. Andrew David, NOAA/NMFS/SEFSC Panama City; Lance Horn, UNCW/NURC - Phantom II ROV operator; 4. Derek Keats from Johannesburg, South Africa, CC BY SA 2.0; P.11: 7. Olympic Coast National Marine Sanctuary, Public Domain; P.13: Sanjay Acharya, CC BY SA 4.0; P.15: Top, Philippe Bourjon, CC By SA 4.0; Bottom, Public Domain; P.17: 1. Smart Destinations, CC BY SA 2.0; 2. NOAA Okeanos Explorer Program, Gulf of Mexico 2012 Expedition; 3. NOAA/NOS/NMS/FGBNMS, National Marine Sanctuaries Media Library; 4. Photographer: Julie Bedford, NOAA PA; 5. NOAA Okeanos Explorer Program, 2016 Deepwater Exploration of the Marianas, Leg 1

If you have walked on a beach, you might have been lucky enough to find a **sea star** in a pool of ocean water.

I want to find a sea star!

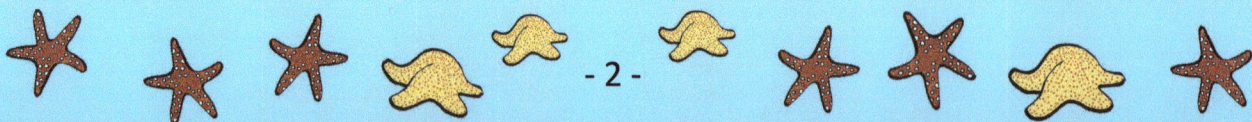

Sea stars are also called **starfish**, but they are not actually fish.

They don't look fishy.

Sea stars are in a
group of animals called
echinoderms.
The word echinoderm
means spiny skin.

I like fur better.

1

Sea stars and other echinoderms are found only in the salty waters of the oceans.

I want to swim in the ocean!

Me too!

Me three!

2

Most sea stars have five arms, which gives them a star-like shape.

Some have more arms. One kind of
sea star can have as many as 24 arms!

Sea stars have a two-sided flat body. The mouth and lots of little **tube feet** are found on the underside.

How would you keep track of that many feet?

Mouth

Tube feet

Underside
of a sea star

A sea star's feet are small hollow tubes that stick out from the underside of the body. Tube feet are used for movement and to grip food.

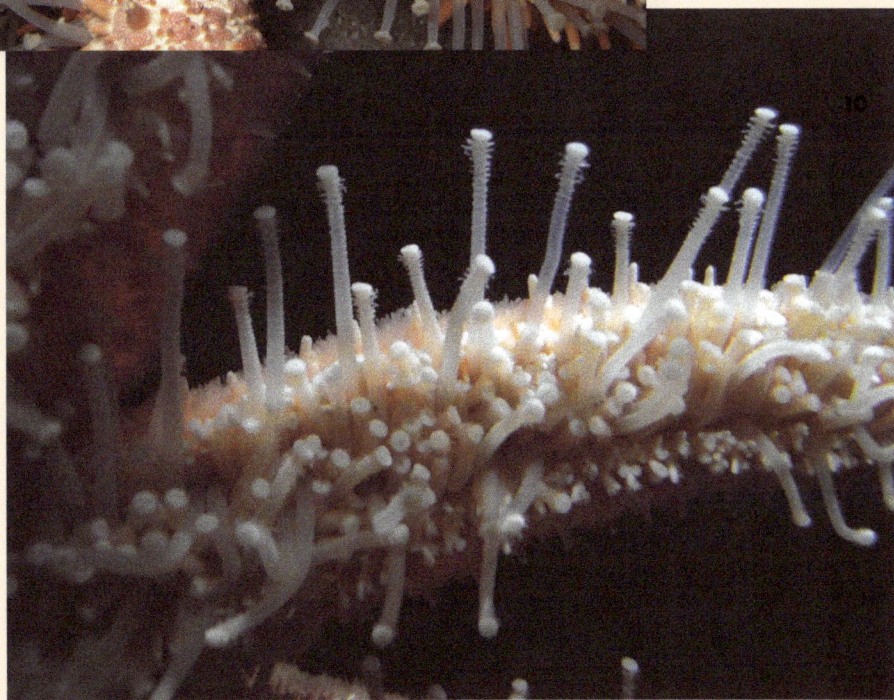

Sea stars come in many different sizes and colors.

Sea stars need to be observed while they are in the water. They die quickly when taken out of water.

Did you know that a sea star has a tiny eye at the end of each arm?

When you look at a sea star, it may be looking back at you!

That sea star is looking at me!

Hello up there! I see you!

How to say science words

echinoderm   (i-KIY-nuh-duhrm)

science   (SIY-uhns)

sea star   (SEE stahr)

spiny   (SPIY-nee)

starfish   (STAHR-fish)

tube feet   (TOOB FEET)

www.ingramcontent.com/pod-product-compliance
Lightning Source LLC
Chambersburg PA
CBHW040152200326
41520CB00028B/7576